D0778556

# The Story of
# IRON

Karen Fitzgerald

A FIRST BOOK

Franklin Watts
A Division of Grolier Publishing
New York London Hong Kong Sydney
Danbury, Connecticut

*To my super brother, Mike*

Chemical consultant: Geoffrey Buckwalter, Ph.D.

Library photo research: Chemical Heritage Foundation, Philadelphia, PA

Cover and interior design by Robin Hessel Hoffmann

Photographs ©: American Museum of Natural History: 11 (Courtesy Department of Library Services), 10 (P. Hollembeak); Art Resource: 15, 38 (Erich Lessing); Comstock: cover center, 36, 25 (Bob Pizaro), 7 (R. Michael Stuckey), 16 (Michael S. Thompson), 43 top (Mike & Carol Werner); Corbis-Bettmann: 45; Fundamental Photos: 41 (Paul Silverman); North Wind Picture Archives: 13, 39; Photo Researchers: cover bottom left (Choromosohm/J. Sohm), 51 (Ken Edward/SS), 54 (David R. Frazier), cover top left (Jan Halaska), 33 (Bob Krueger), 20 (David Nunuk/SPL), cover top right (Alfred Pasieka/SPL), 23 (Charles D. Winters); Photofest: 6; Phototake: 47 (David Bishop), 43 bottom (NASA); Tony Stone Images: 34 (Mark Burnside), 17 (John Lund).

Library of Congress Cataloging-in-Publication Data

Fitzgerald, Karen.
The story of iron / by Karen Fitzgerald
p. cm. — (A First book)
Includes bibliographical references and index.
Summary: Explores the history of the chemical element iron and explains its chemistry and its importance in our lives.
ISBN 0-531-20270-4
1. Iron—Juvenile literature. [1. Iron.] I. Title. II. Series.
QD181.F4F58 1997
669'.141—dc20 96-31541
CIP AC


Printed in the United States of America.
1 2 3 4 5 6 7 8 9 10 R 06 05 04 03 02 01 00 99 98 97

# Contents

# Chapter 1

# THE SUPERMETAL

To most people, *iron* is just a dull, gray metal. It feels cold to the touch, and it turns downright ugly when it rusts. Why would anyone want to read a whole book on iron? Well, iron is not as boring as you think. Lois Lane thought Clark Kent was boring, too, but he actually had many incredible powers. Underneath his glasses and gray suit, he was Superman!

If you take some time to get to know iron, you will find that it is a supermetal with great powers. First of all, it is one of the few substances on Earth that can become magnetic. And although iron is soft and weak, it can transform into one of the strongest materials on Earth. By adding a little bit of an ingredient called *carbon*, it becomes

*Superman transforms from a weakling into the Man of Steel, much like soft iron turns into hard steel.*

steel. It is like Clark Kent putting on his costume and turning into the Man of Steel. Suddenly he can leap tall buildings in a single bound and is more powerful than a locomotive!

Iron gives strength to tall buildings, bridges, and machines. It has built great cities and industries. And it has

*Steel beams provide the strength needed to build skyscrapers.*

made our lives easier. Many appliances and tools operate as a result of the magnetism created by iron. Iron, you see, does a lot of the terribly hard work of the world. It has been our most important metal for 3,000 years, ever since the beginning of the Iron Age. And iron is still the most widely used metal today.

Iron makes our bodies strong, too. Just a tiny amount in our blood carries oxygen to all parts of our bodies, giving us energy. If we don't have enough iron, we feel tired and weak. What is more, iron is an important part of Earth itself. At the center of our planet is a great core of liquid iron that makes Earth a giant magnet with a north and a south pole. A weak magnetism from the core is around us—and inside us—at all times. This magnetism causes the steel needle of a compass to point north, helping people get where they want to go.

So you see, iron is much more than just a boring, gray metal. It is very exciting in its own way. Although this supermetal was discovered more than 6,000 years ago, only in the last 200 years have scientists learned what gives it its amazing powers. In this short time, they have gotten to know the supermetal very well. They now know what makes iron strong and magnetic and how it strengthens our blood.

If you read on, you can get to know iron, too. You will find that it has done many incredible things for our world. But you will also find that it has caused a lot of death and destruction. By the time you finish this book, you will know much of what scientists know about iron and all the ways it has affected our lives. This is the story of iron.

# THE IRON AGE

The first iron that people ever saw came from outer space, historians believe. Ancient people discovered iron in *meteorites*, large rocks that fell to Earth after traveling through space. Oddly enough, Superman also came to Earth from outer space when he was a baby!

Because iron came from the sky, many early people thought iron must have come from heaven. They called it "heaven metal" or "celestial stone." The ancient Egyptians put little beads of iron that they got from meteorites in their jewelry. They mounted the iron beads in gold, which they prized above all other metals on Earth. It's as if the Egyptians realized how important iron would become to civilization.

*This 15-ton iron meteorite fell in the Willamette Valley in Oregon.*

The oldest known iron on Earth is in Egyptian jewelry from around 4000 B.C. We do not make jewelry with iron today because as time passes, it rusts and wears away. But the iron beads from meteorites did not rust away because they contained a small amount of another metal called *nickel*.

Some meteorites are made entirely of iron and nickel. They are called irons. The heaviest meteorite ever found was a 66-ton iron meteorite that landed in Africa. It was discovered in the early 1900s on the Hoba West farm in what is now Namibia.

Early people used meteoritic iron to make tools. Because the iron was mixed with nickel, the iron was as hard as steel. If it had been pure iron, it would have been too soft to make good tools. The iron from meteorites was harder than anything that had ever been seen at the time. No wonder early people thought it was heaven-sent!

*The Hoba West iron meteorite, the largest meteorite ever discovered, was found buried in the ground in Africa.*

The only problem with this iron was that it was very difficult to hammer into shape. Another metal called *bronze* could be melted and poured into any shape. So Egyptians and other ancient peoples made most of their tools from bronze beginning about 3000 B.C. We call this period the Bronze Age.

Eventually, people discovered that they could produce iron by heating a certain kind of rock in a very hot fire, a process called *smelting*. The rock became known as *iron ore*. When iron ore was smelted, a soft, spongy mass of iron oozed out. The iron mass was called a *bloom*, from a word meaning "lump."

## THE DAWN OF A NEW AGE

While it was much easier to make tools from the bloom than from meteoritic iron, there were still problems. The bloom had to be pounded to get rid of a grimy material called *slag*. Then it had to be heated and pounded again. And after all that work, sometimes it wasn't even as strong as bronze. But in about 1500 B.C., people living in the Middle East discovered the secret of making a very hard iron. After repeatedly hammering and heating the iron in a certain way, they dunked it in water or oil to cool it quickly. What came out was steel.

Because steel is stronger than bronze, these people began using iron instead of bronze to make weapons and farm tools. A Middle Eastern tribe called the Hittites kept

*Early blacksmiths had to repeatedly heat and pound iron before it could be formed into tools.*

the secret of making steel to themselves for a long time. But when they were conquered by European invaders in about 1200 B.C., the secret spread to Europe, India, and Asia. That's when the Iron Age began.

Back then, no one knew why steel was harder than other kinds of iron. Most people thought steel must be a pure form of iron. Today we know the opposite is true: steel is iron plus a small amount of carbon, which is the same thing as charcoal. Iron smelters from long ago didn't know it, but iron absorbed carbon from the charcoal fire.

As iron became more important during the Iron Age, it began to get a bad reputation. It was being used to make weapons, and these iron spears, arrows, and swords killed many more people than bronze weapons. A Roman author named Pliny the Elder wrote, "I look upon iron as the most deadly fruit of human ingenuity. For to bring death to men more quickly we have given wings to iron and taught it to fly."

One group of people who used iron very successfully to wage war was the Celts. After becoming quite skilled at making steel tools, they conquered most of Europe by about 500 B.C. with their superior steel swords. The Celts helped the Iron Age take hold across Europe.

To the people in the conquered lands, the Celts were barbarians. The bronze-using cultures looked down on iron as the metal of common, uncivilized people. Iron, historians believe, may have inspired democracy by giving ordinary people an advantage for the first time over the ruling classes. For centuries, bronze weapons had helped rulers control their subjects. The ores needed to make bronze were hard to find, but anybody could get iron ore because there was plenty of it and it was easy to mine.

Since the most common iron ore was reddish, it was easy to spot. The Greeks called the red iron ore *hematite*, or "bloodstone," because it looks like dried blood. Another common iron ore is gray or black. It is called *magnetite* because it is magnetic.

*These now rusted iron tools were made by the Celts about 500 B.C.*

*Hematite, a red iron ore, is often found close to the surface of Earth.*

Many ancient cultures considered iron an inferior metal because it rusts easily and contains a lot of unwanted slag after smelting. It didn't appear to be as pure as gold, silver, or copper—the primary metal in bronze.

## EARTH AND RUST

Ancient Greek thinkers such as Aristotle wondered why iron rusts so easily. Because rust is red and earthy like hematite, he thought rust must be a kind of earth. Aristotle decided that iron must be made of this red earth and water.

Aristotle believed that everything in the world was made from four basic *elements*—earth, water, fire, and air. All metals were made of earth and water. According to his theory, iron rusts so easily because it contains more earth than other metals do. Since gold doesn't corrode at all, Aristotle thought it was the only metal made of pure water.

*The ancient Greeks noticed that rust looks a lot like hematite.*

It was not long before other Greeks got the idea that they could change iron and other "inferior" metals into gold by somehow getting rid of the earth they contained. The attempt to do that is part of an ancient science called *alchemy*. Alchemists mixed and heated iron with many different substances, hoping to transform it into gold. Often, the substances reacted with each other and formed completely new substances. It looked as if the iron was changing into something else, but it never changed into gold.

Even so, the alchemists learned a lot about how chemicals react with one another. By the 1700s, the study of chemistry became more important than making gold. One chemical reaction that received a lot of attention took place when iron was roasted in fire. Iron turned completely to red earth, or rust.

Most scientists thought the fire must have removed something from the iron, leaving behind only earth. But the strange thing was that the red earth weighed more than the original iron. If something was removed from the iron, shouldn't the earth weigh less than the iron?

A French chemist named Antoine Lavoisier wondered about that. He knew that other metals also weighed less than their "earths." By 1785, he realized that scientists had things backward. Iron is one of the elements in the red earth, not the other way around. The heat of the fire had caused iron to combine chemically with oxygen from the air to form a compound, a substance containing more than one element.

Lavoisier realized that earth is not an element at all. Other experiments convinced Lavoisier that fire and water are not elements either. Water is a compound of oxygen and hydrogen, and fire is a chemical reaction in which oxygen from the air combines with the burning substance.

Lavoisier's discoveries proved that the theory of the four elements was wrong. He realized that there is a whole world of other elements, and one of them is iron. He included it with more than 30 substances that he suspected were elements because they had not been broken down into anything else. Lavoisier called iron *fer*, which comes from the Latin word for iron.

It took a while for other chemists to accept Lavoisier's new theory. But when they did, they began discovering many new elements. To date, chemists have discovered 92 naturally occurring elements, and they have artificially generated at least 20 more. Lavoisier's work brought about a major change in chemistry, steering it toward the science it is today.

Chemists began to understand what was really happening in chemical reactions, such as the smelting of iron ore. Iron ore is a compound that separates into its elements, iron and oxygen, when heated to a high temperature. The iron flows out, while the oxygen combines with the carbon in charcoal to form a gas called *carbon dioxide*, which escapes into the air.

Chemists started naming compounds according to the elements they contain. Today chemists call hematite

*In the 1700s scientists did not believe that chunks of iron or rock could fall to Earth from space.*

*iron oxide* to indicate that it contains iron and oxygen. Magnetite is also an iron oxide.

Strangely enough, during this time of incredible progress for chemistry, the study of meteorites had a great setback. Many scientists, including Lavoisier, believed it was impossible for chunks of iron or rock to fall to Earth from outer space. They thought meteorites were actually formed on earth, and they made fun of people who claimed the rocks had come from outer space. In 1790 the French Academy of Sciences announced that it would no longer investigate reports of stones falling from the sky. But when a meteorite fell in France in 1803, these scientists realized they were wrong.

So you see, in the 1800s, the truth came out about iron. Scientists recognized not only that it is an element, but that it sometimes comes to Earth from outer space. They would also discover that iron has powers they never dreamed of before. Iron, revealed as a superhero at last, would change the world.

# A VERY UNCOMMON METAL

In the new chemistry, iron could no longer be considered inferior to other metals. It was an element just like the rest, with its own unique characteristics, or *properties*. Nineteenth-century chemists began to wonder what exactly gives iron and other elements their special properties.

Like detectives tracking down a mystery, chemists tried to get clues by carefully watching how the elements behave. They mixed and heated elements together to see if they would react and form compounds. Lavoisier, for instance, studied how iron combines with oxygen. He set some iron on fire in pure oxygen and discovered that a black oxide formed. But then when he heated the same black powder in air, it turned brown. It had changed to a

*Lavoisier found that burning bits of iron in pure oxygen produced a beautiful spectacle—and a black iron oxide.*

different compound—a kind of hematite—as a result of absorbing more oxygen.

This test showed that the only difference between iron oxides such as magnetite and hematite is the amount of iron and oxygen each contains. Today we know that red hematite always contains 70 percent iron and 29 percent oxygen, while magnetite always contains 72 percent iron and 27 percent oxygen.

Lavoisier had noticed that many other compounds seemed to contain a constant proportion of elements, too, but he did not know why. In the early 1800s, an English chemist named John Dalton solved the mystery. He realized that elements combine in constant proportions because they are made of particles called *atoms*. The atoms from different elements bond together in groups to form compounds. As a result, each compound contains a unique group of atoms that is repeated throughout the material. These groups are called *molecules*.

A molecule of red hematite contains two iron atoms bonded to three oxygen atoms, while a molecule of magnetite contains three iron atoms attached to four oxygen atoms. Compounds can be identified by the atoms in their molecules. The chemical symbol for the iron atom is Fe, which comes from *fer*; and for the oxygen atom, it is O. So $Fe_2O_3$ represents hematite, and $Fe_3O_4$ is the formula for magnetite.

Dalton believed that the atoms of each element are somehow different from those of every other element and that these differences are responsible for each element's

*Hematite, on the left, and magnetite, on the right, are both compounds of iron and oxygen.*

unique properties. He guessed that an iron atom and an oxygen atom, for instance, are different in size and weight.

Although he could not weigh atoms because they are too tiny, Dalton was able to estimate the weight of each atom. He assumed that an atom of hydrogen—the lightest element—had a weight of 1 and calculated that an atom of iron weighs 50. Today we know it weighs 56 times more than a hydrogen atom.

Why does an iron atom weigh so much more than a hydrogen atom? It has to do with what's inside it, chemists later discovered. All atoms contain particles called *protons* and *electrons*. A hydrogen atom has one proton with one electron flying around it. Because the proton is about 1,800 times heavier than the electron, most of the weight of the atom comes from the proton.

An iron atom has 26 protons gathered in its center, or *nucleus*, plus 30 more particles weighing slightly more than protons. These particles are called *neutrons*. With 56 protons and neutrons, the iron atom weighs 56 times more than a hydrogen atom.

There are also 26 electrons flying around the iron nucleus. They may not weigh much, but they are important because they determine how iron interacts with other elements.

## WHAT BONDS ARE MADE OF

The atoms in a molecule do not simply mix together like marbles in a bowl; they actually attach to one another. Elements bond only if the electrons in their atoms fit together well. You see, electrons tend to gather in groups of two or eight. It is as if electrons lived in homes that can hold either two or eight electrons. Each "home" is like a ring located at a different distance from the nucleus. Only two electrons can live in the ring closest to the nucleus. But the second ring can hold a family of eight electrons.

An oxygen atom has eight electrons—two in its first family and six in the second. Because electrons seem to be happy only if their homes are completely filled, oxygen would like to bond with other atoms that have two electrons to spare. Hydrogen atoms are good partners for oxygen because each one has a single lonely electron. Two hydrogen electrons gladly fill the empty spots in oxygen's outer ring. The result is a chemical bond between one oxygen atom and two hydrogen atoms, which is $H_2O$, or water.

Chemical bonds with iron are more difficult to explain because its outer family of electrons is rather odd. The first 18 electrons in an iron atom are part of three complete families: two electrons in the first ring and eight in each of the next two rings. You might think that its eight remaining electrons would form a complete family, too, but they don't.

It's as if these electrons lived in a great mansion with an added wing of bedrooms. The main part of the house can hold eight electrons just as in the previous rings, but the wing has space for ten more electrons. Strangely enough, only two of iron's outer electrons stay in the main part of the house, while the remaining six electrons stay in the wing. The electrons in the wing are called *d* electrons.

Since it would be difficult to find atoms that can fill the openings in iron's outer family, iron donates electrons to atoms that need electrons. But iron usually supplies only two or three electrons, and six at the most. Chemists believe iron gives up just a few electrons because some of the *d* electrons are attracted too strongly to the iron nucleus to join the electrons of another atom.

Iron gives two electrons to an oxygen atom to form

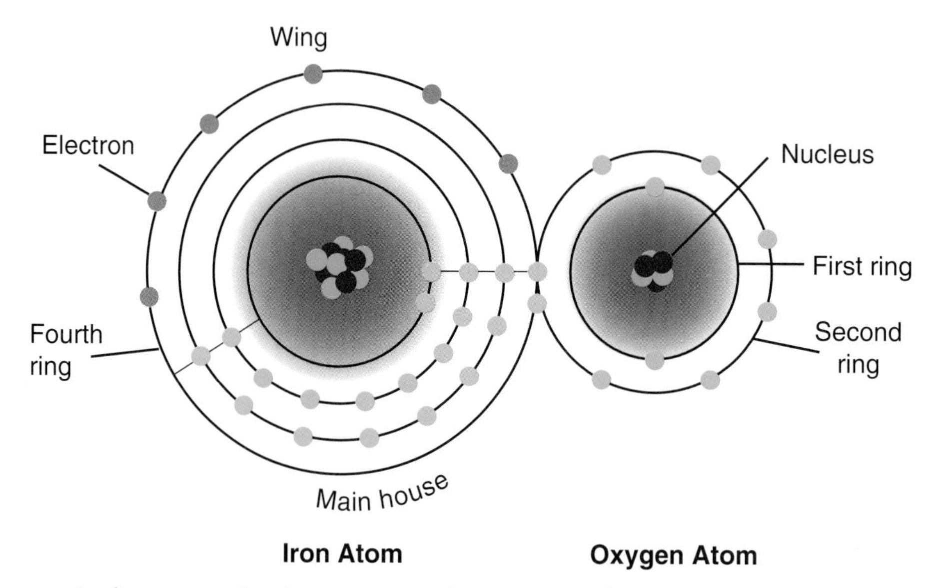


*In ferrous oxide, the iron atom donates two electrons to the oxgyen atom so that it has a complete outer family of eight electrons.*

*ferrous oxide*, which has a formula of FeO. This compound is not very common because it changes easily into other oxides. In hematite, $Fe_2O_3$, each iron atom donates three electrons, to give a total of six electrons to three oxygen atoms. Hematite is also called *ferric oxide*.

As you can see, iron is not afraid to break the rules that most elements follow when they form compounds. Magnetite, $Fe_3O_4$, is another good example. One iron atom in the molecule donates two electrons, while the other two iron atoms donate three electrons apiece. That allows them to supply four oxygen atoms with a total of eight electrons.

## THE WILD ONES

Iron and other elements with a wing partly filled with *d* electrons are called *transition elements*. As shown on the next two pages, they fill the middle of the periodic table of elements and have some similar properties. Like all transition elements, iron is a metal, which means it conducts electricity. All transition elements tend to be strong, shiny, and heavy.

The two electrons in the main part of the outer ring make it possible for iron to conduct electricity. They are like kids who go a bit wild when their parents aren't around. Instead of sticking close to home, they run around in packs with electrons from surrounding iron atoms. If any part of the iron material becomes positively charged because of missing electrons, the electrons race toward it, creating a flow of electricity.

The two conducting electrons also help iron conduct heat very well. They absorb the heat energy and take it to colder parts of the metal. Iron feels cold when you touch it because it quickly carries heat away from your hand. But when an iron object is hotter than your hand, you may be burned because the heat travels so quickly to your hand.

The conducting electrons from all the atoms form a kind of cloud around the atoms even when electricity is not flowing. Light cannot go very far into the metal because the cloud reflects it at the surface. As a result, iron looks shiny to us.

# Periodic Table

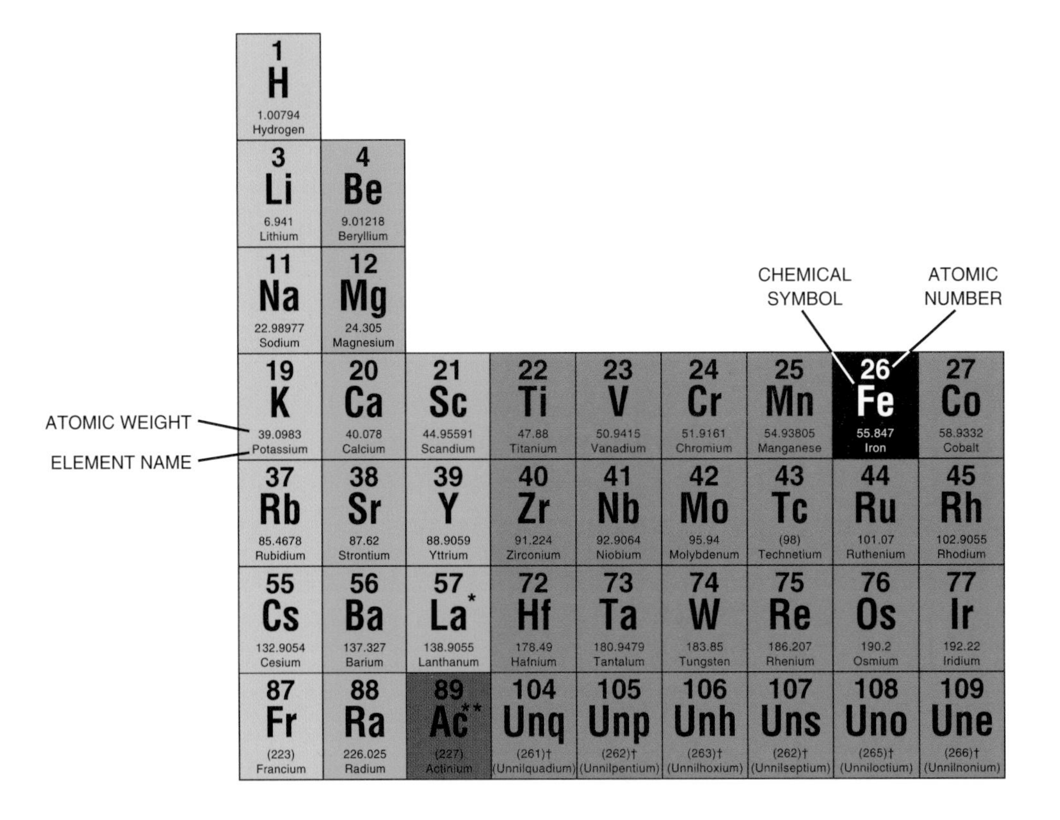


*Elements are organized in this table by* atomic number, *the number of protons the atom contains. Iron is one of the transition elements (in blue), which tend to be heavy metals.*

| 58 | 59 | 60 | 61 | 62 |
|---|---|---|---|---|
| Ce | Pr | Nd | Pm | Sm |
| 140.115 | 140.9077 | 144.24 | (145) | 150.36 |
| Cerium | Praseodymium | Neodymium | Promethium | Samarium |

| 90 | 91 | 92 | 93 | 94 |
|---|---|---|---|---|
| Th | Pa | U | Np | Pu |
| 232.0381 | 231.0359 | 238.029 | 237.048 | (244) |
| Thorium | Protactinium | Uranium | Neptunium | Plutonium |

# *of the Elements*

| | | | | | | | | |
|---|---|---|---|---|---|---|---|---|
| | | | | | | | | 2 **He** 4.00260 Helium |
| | | | 5 **B** 10.811 Boron | 6 **C** 12.011 Carbon | 7 **N** 14.067 Nitrogen | 8 **O** 15.994 Oxygen | 9 **F** 18.998403 Florine | 10 **Ne** 20.1797 Neon |
| | | | 13 **Al** 26.96154 Aluminum | 14 **Si** 28.0855 Silicon | 15 **P** 30.973762 Phosphorous | 16 **S** 32.066 Sulfur | 17 **Cl** 35.4527 Chlorine | 18 **Ar** 39.948 Argon |
| 28 **Ni** 58.693 Nickel | 29 **Cu** 63.546 copper | 30 **Zn** 65.39 Zinc | 31 **Ga** 69.723 Gallium | 32 **Ge** 72.61 Germanium | 33 **As** 72.9216 Arsenic | 34 **Se** 78.96 Selenium | 35 **Br** 79.904 Bromine | 36 **Kr** 83.80 Krypton |
| 46 **Pd** 106.42 Palladium | 47 **Ag** 107.8682 Silver | 48 **Cd** 112.41 Cadmium | 49 **In** 114.82 Indium | 50 **Sn** 118.71 Tin | 51 **Sb** 121.757 Antimony | 52 **Te** 127.60 Tellurium | 53 **I** 126.9045 Iodine | 54 **Xe** 131.29 Xenon |
| 78 **Pt** 195.08 Platinum | 79 **Au** 196.9665 Gold | 80 **Hg** 200.59 Mercury | 81 **Ti** 204.383 Thallium | 82 **Pb** 207.2 Lead | 83 **Bi** 208.9804 Bismuth | 84 **Po** (209) Polonium | 85 **At** (210) Astatine | 86 **Rn** (222) Radon |

| | | | | | | | | |
|---|---|---|---|---|---|---|---|---|
| 63 **Eu** 151.965 Europium | 64 **Gd** 157.25 Gadolinium | 65 **Tb** 158.9253 Terbium | 66 **Dy** 162.50 Dysprosium | 67 **Ho** 164.9303 Holmium | 68 **Er** 167.26 Erbium | 69 **Tm** 168.9342 Thulium | 70 **Yb** 173.04 Ytterbium | 71 **Lu** 174.967 Lutetium |

| | | | | | | | | |
|---|---|---|---|---|---|---|---|---|
| 95 **Am** (243) Americium | 96 **Cm** (247) Berkelium | 97 **Bk** (247) Berkelium | 98 **Cf** (251) Californium | 99 **Es** (252) Einsteinium | 100 **Fm** (257) Fermium | 101 **Md** (258) Mendelevium | 102 **No** (259) Nobelium | 103 **Lr** (260) Lawrencium |

The electron cloud also makes iron very dense and heavy. With two outer electrons gone, the iron atoms are able to snuggle together very closely. Iron is soft because the atoms can easily slide around each other. They aren't held in place as they would be if the two bonding electrons formed normal, rigid connections with nearby atoms.

As a result of its softness, iron is easily hammered into different shapes. In other words, it is very *malleable*. It is *ductile* too, which means it can be drawn out to form a long, thin wire. Hammering or working iron in any way can make it harder by shifting the atoms into stiff arrangements. Another way to stiffen the arrangement of iron atoms is to add an element such as carbon. The carbon atoms can strengthen the iron atoms by either mixing in or bonding chemically with them.

Chemists discovered that carbon is the secret of steel's strength in the 1700s. Steel contains a very small amount of carbon—no more than 2 percent. It makes iron strong, but still soft enough to be formed into shapes. After it has been shaped, hot steel is plunged into cold water or oil to stress the atoms so that they harden into position. Small amounts of other elements such as nickel, manganese, and phosphorus can also strengthen iron—and prevent rusting.

Rust comes from a chemical reaction between iron, oxygen, and water. First, water in the air dissolves the iron atoms. Then oxygen reacts with the iron, forming ferric oxide. Rust is a type of ferric oxide that includes water in

*The brownish red rust on this old car formed after years of exposure to oxygen in the air and rain water.*

each molecule. As a result, it is brownish red instead of the bright red of pure ferric oxide.

The added elements in steel protect against rust by surrounding the iron atoms, making it more difficult for iron to react with water and oxygen. One element, chromium, prevents *corrosion* altogether. Steel containing

*Stainless steel does not rust because the iron is protected by chromium atoms.*

chromium is called "stainless steel" because it doesn't change color over time.

For strengthening iron, however, carbon is the element of choice. It transforms iron from a soft, ordinary metal into the metal of skyscrapers and machines.

# FORGING INDUSTRY

Metropolis, Superman's hometown, was a city of skyscrapers and bustling industry. It is a good example of the amazing things steel has done for people. The invention of machines and improvements in steel making in the 1800s brought about rapid changes that produced our modern world.

Progress in making steel started long ago when smelters found a way to produce liquid iron, instead of the spongy iron bloom. Liquid iron could be cast, or poured into molds; it did not have to be hammered, or "worked," like the bloom. Chinese smelters were making *cast iron* as early as 400 B.C., but Europeans didn't learn to do it until much later.

*Smelters eventually began producing liquid iron because it can be poured into a mold of the desired shape.*

One secret of making cast iron was to force a lot of air into the smelting fire so that it would burn hotter. By the 1200s, iron-smelting furnaces in Spain had mechanisms that automatically forced air to flow across the burning charcoal. The iron was softer, but still not liquid.

Next, European furnaces were built with tall stacks. Charcoal and iron ore were fed into the top of the stack, and the ore was smelted as it slowly moved down the stack. Because the stacks were so tall, the iron stayed in contact with the burning charcoal for a long time and absorbed a lot of carbon. The more carbon in the iron, the lower its melting point becomes. So the iron came out much softer at the bottom of the furnace. Eventually, people built stacks so tall that the iron came out completely melted.

Europeans could then mold the iron into the shapes they wanted. The only problem was that the iron was brittle because it contained so much carbon. When iron contains more than about 3 percent carbon, it becomes so stiff that it breaks easily. Fortunately, Europeans found a way to reduce the carbon to less than 2 percent, which converted the cast iron to steel. They mixed the liquid iron with hematite. The oxygen in the hematite combined with some of the carbon and formed carbon dioxide, which rose into the air as a gas.

In the 1500s, the tall furnace developed into what is known as the *blast furnace*. It included devices at the bottom that blasted air up into the charcoal and iron ore at high speeds. Then in the early 1800s, charcoal was replaced

*Iron ore from the mine, at right, is fed into the top of a tall smelter that produces cast iron. The water wheel keeps air blowing into the fire. (From a 1580 painting by Lucas van Valckenborch)*

*The Brooklyn Bridge, built in 1883, was the first bridge to be suspended by steel cables.*

with another carbon fuel called coke, and iron became much cheaper. It was no longer used just in tools, but in buildings, bridges, and boats. Demand for iron skyrocketed. It became even cheaper when a faster way of removing carbon from cast iron was discovered in 1856; the molten iron was blasted with pure oxygen.

Today the iron industry still uses blast furnaces, but they are much more sophisticated. One blast furnace about 100 feet (33 meters) tall can produce as much as 10,000 tons of iron a day! Finding enough iron ore to feed all the iron furnaces in the world would be a problem if iron were not the second most plentiful metal in Earth's crust. The world's largest iron ore mine, which is in Australia, produces 40 million tons of hematite every year.

One of the greatest impacts of the steel boom in the 1800s was in the machines that drove the Industrial Revolution. The machines relied on the magnetism of steel as well as its strength. Modern machines were invented only after scientists began to understand magnetism.

## THE MYSTERIOUS ATTRACTION

People had tried to uncover the secret of magnetism since ancient times. Thales, a Greek philosopher from the sixth century B.C., was one of the first to wonder how a magnet works. The first magnets were rocks containing magnetite, and they were called *lodestones* back then. Thales guessed that a lodestone could attract iron because it had a soul.

Others after Thales thought lodestones must be living things. If a lodestone was hung in the air by a string, the stone rotated until it lined up with the north and south directions. Travelers began taking lodestones with them to help find their way. The Chinese found that rubbing a

*A lodestone, which is made of magnetite, attracts the magnetized steel needle of a compass.*

steel needle on a lodestone made the needle magnetic too. They began making compasses with a magnetized needle that always pointed south, and Europeans later made compasses that pointed north.

But no one knew how compasses worked. People eventually discovered that every magnet, whether it is a lodestone or a steel needle, has a north pole at one end and a south pole at the opposite end. A magnet's north pole is attracted to the south poles of all other magnets, and its south pole is attracted to the north poles of other magnets. Magnets point north, some guessed, because there is a big magnet in the heavens near the North Star.

In the 1500s, an English doctor named William Gilbert noticed that as a compass needle points north, it tilts very slightly down toward the ground. He guessed that Earth itself is a giant lodestone, pulling the needle downward. Today we know that he was right. Earth contains a great core of molten iron that acts as a magnet with poles near the North Pole and the South Pole. That is why the magnetic needles in compasses always point north.

But how does magnetism work over such great distances? In the early 1800s, an English physicist named Michael Faraday tried sprinkling iron filings on a piece of paper above a magnet. To his surprise, the filings lined up along curves extending from one pole of the magnet to the other. Faraday visualized these lines of magnetic force filling the space around a magnet, creating what he called a *magnetic field*. Anything in this field experiences magnetism.

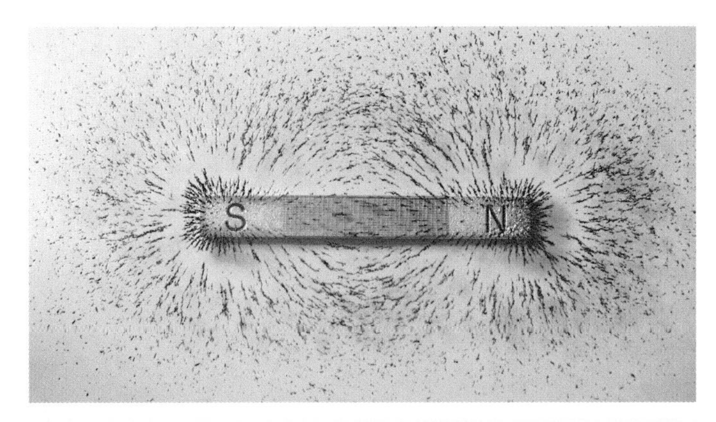


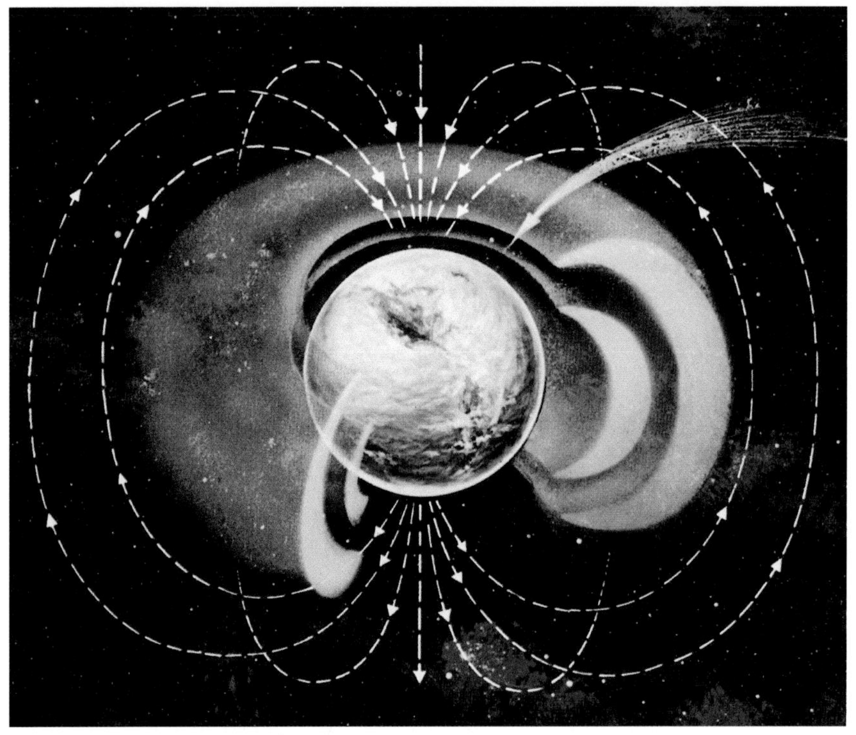

*Iron filings outline the shape of the magnetic field surrounding a magnet. A similar magnetic field surrounds Earth.*

Not long before Faraday's work, scientists had been amazed to discover that electricity flowing in a wire produces magnetism. Two wires carrying electricity in opposite directions attracted each other just like magnets! And when wire was wrapped around an iron bar, the bar became a powerful *electromagnet* capable of lifting 20 times its own weight whenever electricity flowed through the wire. In an instant, electricity changed iron from a weakling to a supermetal.

Faraday suspected that magnetism could produce electricity too. In 1830 he showed that electricity automatically flows in a wire whenever it is in a magnetic field that is changing in strength or direction. As a result of this discovery, he invented the electric generator. It generates electricity by rotating a circular coil of wire through a magnetic field so that the magnetism in the wire constantly changes.

Powered by steam or water, the electric generator offered a continuous supply of electricity for the first time. Today, huge generators at power plants produce electricity for homes and offices. A small generator under the hood of a car creates the electricity for the headlights and radio. So you see, the electric generator changed people's lives forever. Lighting homes became easier, and electric appliances, tools, and other devices became possible.

Many appliances and tools contain an electric motor, which is really just a generator working in reverse. When electricity is applied to the coil of wire, it rotates because the coil is in a magnetic field. Anything connected to the coil, such as a washer, dryer, or lawn mower blade, spins too. Without generators and motors, we would not have most of our modern conveniences.

Motors and generators became important parts of machines that brought about an industrial revolution in the 1800s. An explosion in industry took place as factories began producing cheaper products. Machines continue to manufacture the great variety of products in stores today.

*Coils of wire rotating in a magnetic field generated electricity in an early power plant.*

And iron is the heart of motors, generators, and electromagnets. In these devices, the electric wire is wrapped around a core of iron to channel and strengthen the magnetic field the wire generates. Iron is ideal for this purpose because it becomes magnetized only temporarily. When the electricity stops, the iron goes back to being ordinary and nonmagnetic—a lot like Clark Kent! Motors and generators may also contain *permanent magnets*, which are always magnetic, like the magnets on a refrigerator. Permanent magnets usually contain iron in the form of steel, iron oxides, or a mixture of iron, aluminum, nickel, and cobalt.

Magnetic tapes and disks record computer data, sound, and video on semipermanent magnetic material. This material is a thin layer of iron oxide that can be magnetized when a magnetic recording head moves over it. Because the iron oxide is not permanently magnetic, the data can be erased or written over when the magnetic head passes over again.

Nickel and cobalt, as well as a few other elements, can become magnetic, but iron makes the strongest magnet of all the elements. Why is iron so good at becoming magnetic? It has to do with the *d* electrons in an iron atom. Scientists have discovered that magnetism comes from the spinning of electrons inside atoms. All electrons spin, but in most atoms, half spin in one direction and half spin in the other, canceling out the magnetism. Because of the *d* electrons in an iron atom, however, more electrons spin in one direction than the other, so each iron atom is like a

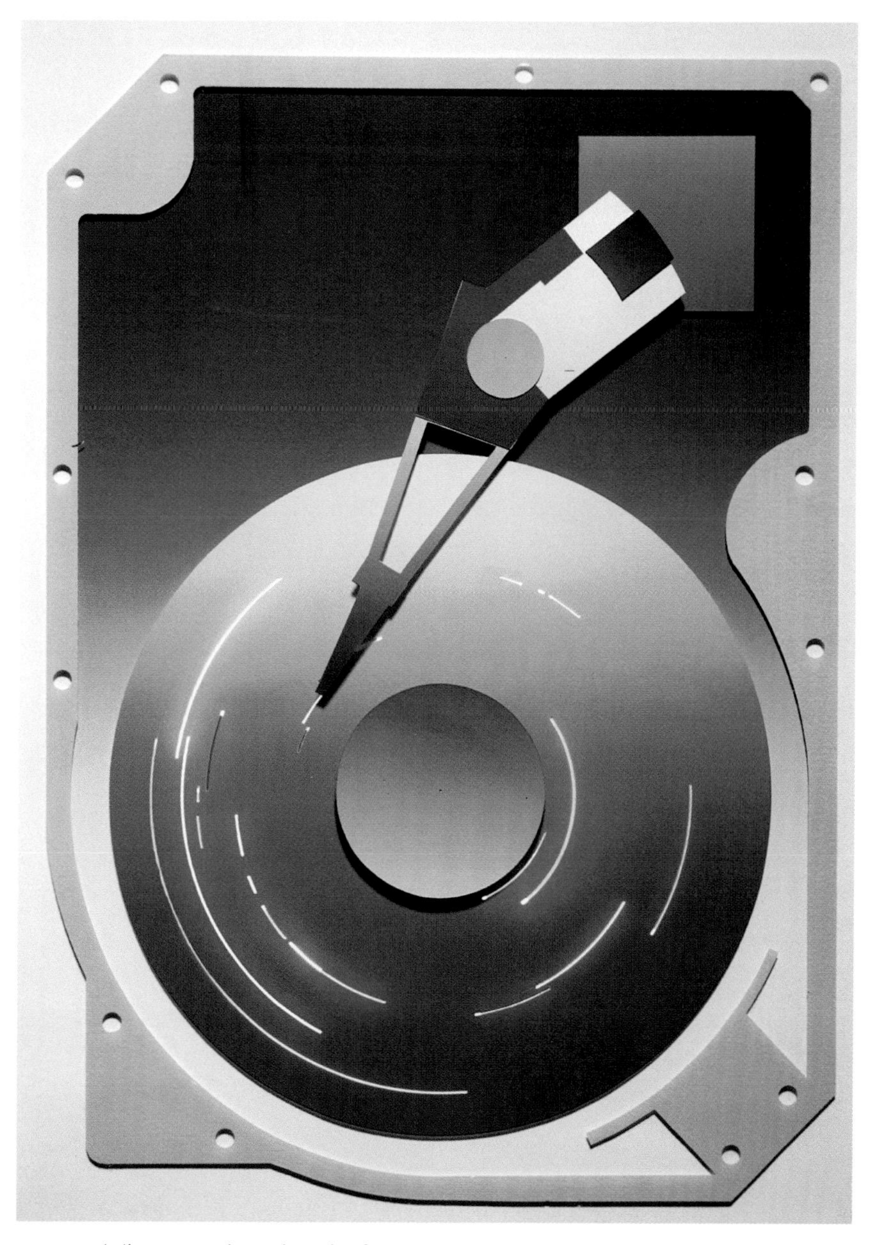

*The recording head of a computer disk drive magnetizes a film of iron oxide in tracks along the disk.*

tiny magnet. When a piece of iron is in a magnetic field, all the atomic magnets line up in the same direction, creating one giant magnet. This property is called *ferromagnetism*.

So you see, magnetism transforms iron into a supermetal in a very different way than carbon does. The intimate connection between electricity and magnetism made it possible for iron to help us even more in our everyday tasks. In many ways, iron is responsible for much of the technology we have today. As if that weren't enough, the supermetal also helps us with very important tasks in our body.

# IRON MAN

Iron is so cold and gray that it seems the very opposite of a living thing. How could the material of machines, buildings, and bridges possibly be part of people, animals, and plants? It is just one more amazing thing the supermetal can do.

It took people a long time to connect iron with living things. Not until the beginning of the 1700s did scientists begin seeing clues that plants and animals contain iron. First, it was found in the ashes of plants that had been burned. Then a French chemist named Louis Lemery discovered iron in soil and guessed that it is taken up by plants in water. He was right about that.

Today scientists know that plants need iron to make *chlorophyll*, the green material in plants. Although chlorophyll does not contain iron, an iron compound helps make chlorophyll. Plants use chlorophyll to absorb the sunlight they need to make food and to grow.

In 1745, an Italian doctor named Vincent Menghini discovered that animals need iron, too. He burned some dried blood from a dog and found that particles in the ashes were attracted to a magnet. The particles, he realized, had to be iron. He later traced the iron to red blood cells. Eventually scientists discovered that blood is red because of iron. Iron in blood combines with oxygen in the lungs, turning the blood bright red. It is the same chemical reaction that takes place when iron is heated in air and turns to red rust.

Although red blood cells carry only a tiny amount of iron, it is very important. It must carry oxygen to all the cells of our bodies so that we can live. Our muscles, in particular, rely on the oxygen from iron to carry out strenuous tasks.

Iron is part of a complex substance called *hemoglobin* in red blood cells. A hemoglobin molecule contains four groups of atoms called heme, which each have a single iron atom in the center. As a blood cell passes through the lungs, two oxygen atoms attach to each iron atom. Fortunately, blood cells can hold lots of hemoglobin. Because hemoglobin is concentrated in cells, the blood carries 100 times as much oxygen as it would without cells. Strangely

*Blood cells (in the background) contain hemoglobin, which consists of four heme groups (the white atoms) attached to a protein called globin (in red). At the center of each heme group is a red iron atom attached to two blue oxygen atoms.*

enough, this means the percentage of oxygen in the blood is exactly the same as the percentage in the air.

The bond between the iron atom and oxygen atoms is not like most chemical bonds. Its strength depends on how much oxygen is in surrounding tissues. Where there is a lot of oxygen—near the lungs, for example—the iron atom holds the oxygen atoms tightly. But as it travels to other parts of the body, the iron releases the oxygen to body cells. These cells use the oxygen to create the energy needed to keep the body operating.

For such an important substance, iron is only a very small part of the body. In an average person, the iron weighs about as much as five paper clips. Yet if we don't get enough iron, our blood carries less oxygen to our cells and we feel tired. Doctors call this condition anemia, or iron-poor blood.

So you see, iron strengthens not only buildings and bridges, but our bodies, too. To prevent anemia, it is important to eat foods that contain iron. Red meats and leafy green vegetables, such as spinach, are rich in iron. Egg yolks, carrots, fruit, and whole wheat also contain iron.

## IRON OVERLOAD

Iron may be beneficial inside the body, but some people think the magnetism produced by iron outside the body could be harmful. They are concerned because they believe that magnetic fields from electric appliances and power lines that supply electricity to homes and offices may cause cancer or other health problems.

Magnetic fields surround blenders, TVs, computer screens, and power lines, just as they do magnets. Anyone who comes close to them while they are operating is exposed to the magnetism. Our bodies are always exposed to Earth's magnetic field with no ill effects, but it is much weaker than the fields near power lines, appliances, and computers.

Scientists do not believe the magnetism from these technologies is harmful to our bodies, but cells do behave differently in magnetic fields. It's not surprising, since there is electrical activity in cells, and wherever there's electricity there's magnetism.

Perhaps fears about magnetic fields come from the notion that our world has become too dependent on tedhnology. Even if magnetic fields are safe, some people say that all the steel and technology surrounding us now cuts us off from nature. Another complaint is that the environment is suffering from mining so much iron ore. Are we using too much of the technology iron has brought us?

Looking back on everything iron has done for civilization, it is clear that it has helped us immensely. In reaching up toward the heavens in skyscrapers, iron has fulfilled the promise of its ancient name—the heaven metal.

*Large transmission power lines distribute electricity to the smaller power lines to the left. Does the magnetism generated by the transmission lines affect people living nearby?*

By doing most of our hard work, it has given us time to develop our minds and our culture. But after more than 3,000 years of toil and feats of wonder, perhaps it is time to give the supermetal a much deserved rest.

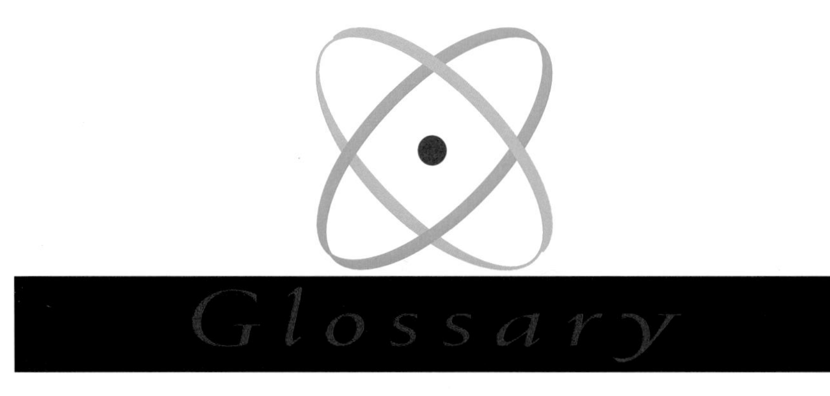

# Glossary

**alchemy**—a practice in which people experimented with chemicals to find a way to change metals such as iron into gold. Some alchemists tried to cure diseases and achieve spiritual perfection.

**anemia**—an illness caused by too little iron or hemoglobin in the blood.

**atom**—a particle that is the smallest piece of an element. Each element has an atom with a unique number of protons and electrons.

**blast furnace**—a furnace for smelting iron ore that blasts air into the fire to keep it burning hot.

**bloom**—a mass of soft iron that can be worked into the desired shape. In ancient times, smelting iron ore produced iron in this semi-solid state.

**bronze**—an alloy, or mixture, of copper and tin.

**carbon**—an element added to iron to make steel. In its pure form, it is usually black, as in charcoal.

**carbon dioxide**—an invisible gas that animals exhale and plants absorb. It is made of carbon and oxygen, and its formula is $CO_2$.

**cast iron**—a kind of iron that results when iron ore is smelted to produce liquid iron, which can then be poured into molds. Its high carbon content makes cast iron brittle, so it cannot be worked like low-carbon steel or iron.

**chlorophyll**—the green substance in plants that absorbs sunlight as part of photosynthesis.

**coke**—a fuel containing carbon made from coal. It is used to make steel.

**compound**—a substance made up of two or more elements chemically bonded together.

**corrosion**—a gradual wearing away.

**ductile**—able to be drawn out or hammered into a desired shape.

**electromagnet**—a piece of iron with wire wrapped around it. It becomes a magnet when electricity flows in the wire.

**electron**—a tiny particle that orbits the nucleus of an atom. It has a negative electrical charge.

**element**—a basic substance that cannot be broken down into any simpler substance. There are 92 elements that occur naturally, and at least 20 more have been generated artificially.

**ferric oxide**—a compound of iron and oxygen found in the mineral hematite. Its formula is $Fe_2O_3$.

**ferromagnetism**—the ability to become magnetized easily.

**ferrous oxide**—a compound of iron and oxygen with the formula FeO.

**hematite**—a red mineral that is an iron ore containing 70 percent iron.

**hemoglobin**—a compound containing iron that carries oxygen in the blood to all parts of the body and returns carbon dioxide to the lungs.

**iron**—an element with twenty-six protons and electrons.

**iron**—a meteorite containing mostly iron, with about 10 percent nickel.

**iron ore**—any of a variety of minerals, or rocks, that contain so much iron that people mine the ore to produce iron. Hematite and magnetite are the most common iron ores.

**iron oxide**—any compound of iron and oxygen. The two most common are $Fe_2O_3$ and $Fe_3O_4$.

**lodestone**—the ancient name for the mineral magnetite.

**magnetic field**—the space around a magnet where there is a magnetic force.

**magnetite**—a black mineral; an iron ore containing 72 percent iron.

**malleable**—capable of being shaped with a hammer or other tool.

**meteorite**—a meteor that falls to Earth. Meteors are pieces that have broken off from comets or asteroids traveling around the solar system.

**molecule**—a group of two or more atoms bonded together, usually forming a compound.

**neutron**—a tiny particle in the nucleus of an atom. It has no charge.

**nickel**—a gray metallic element that is often mixed with other metals to prevent corrosion. The five-cent nickel contains 25 percent nickel and 75 percent copper.

**nucleus**—the central part of an atom.

**permanent magnet**—a magnet that stays magnetized even when it is removed from a magnetic field.

**property**—a quality or behavior that a substance displays.

**proton**—a particle in the nucleus of an atom. It has a positive electrical charge.

**slag**—the scum that forms on the surface of a molten metal. It is usually removed because it contains impurities

**smelting**—a process used to separate a metal from an ore.

**transition element**—an element in the middle of the periodic table. Transition elements are all metals with some *d* electrons in the outer rings of their atoms.

# Sources


Brock, William H. *The Norton History of Chemistry*. New York: W.W. Norton, 1992.

*Dictionary of Scientific Biography*. New York: Scribner, 1970–80.

Forbes, R. J. *Studies in Ancient Technology*, Vol. 9. Leiden, the Netherlands: E. J. Brill, 1964.

Greenwood, N. N., and A. Earnshaw. *Chemistry of the Elements*. Oxford, England: Pergamon Press, 1984.

Hudson, John. *The History of Chemistry*. New York: Routledge, Chapman & Hall, 1992.

Meyer, Herbert W. *A History of Electricity and Magnetism*. Cambridge, Mass.: MIT Press, 1971.

Verschuur, Gerrit L. *Hidden Attraction: The History and Mystery of Magnetism*. New York: Oxford University Press, 1993.

Weeks, Mary Elvira. *Discovery of the Elements*. Easton, PA: Journal of Chemical Education, 1968.

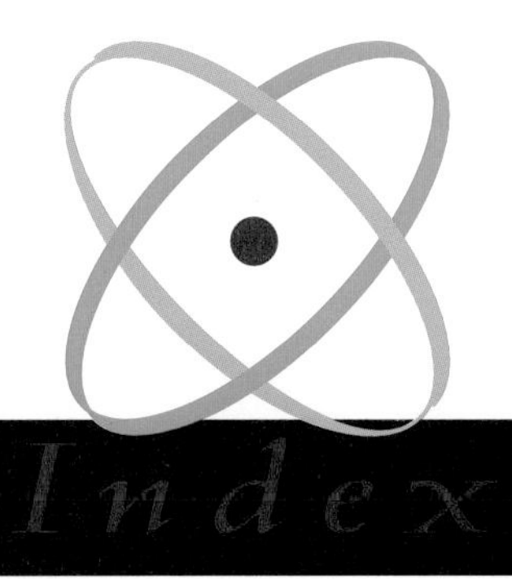

# Index

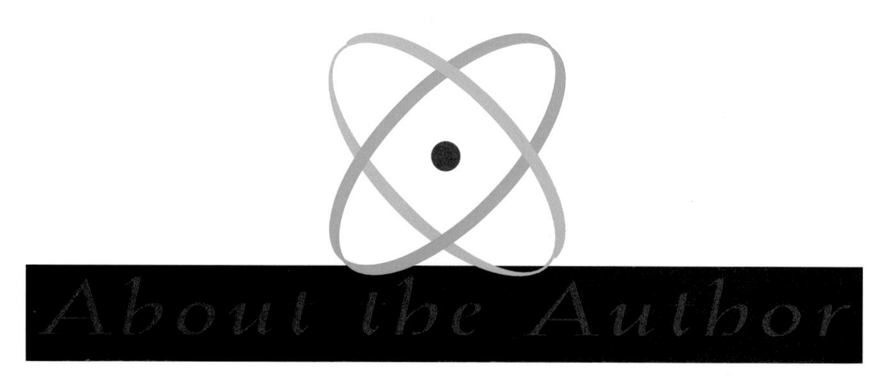

# About the Author



Karen Fitzgerald is a science writer and former Franklin Watts editor. She has also worked as an editor for *The Sciences*, a magazine published by the New York Academy of Sciences, and *Spectrum*, a technology magazine published by the Institute of Electrical and Electronics Engineers. She has written articles for magazines including *Scientific American*, *Omni*, and *Science World*. Her bachelor's degree is in mechanical engineering from the University of Illinois in Urbana, and she has a master's degree in science and environmental reporting from New York University.